# RESUSCITATION:
## key data

Also of interest:

*Key Topics in Anaesthesia*
T.M. Craft and P.M. Upton
Publication date March 1992, revised and reprinted June
1993

ISBN 1 872748 90 2

This book contains essential information on a broad range
of selected topics pertinent to modern anaesthesia.   The
information is presented in a problem-based format and provides a unique revision aid for candidates of the post-gradu-
ate examinations in anaesthesia.

# RESUSCITATION:
## key data

**M. J. A. Parr**
MB BS  MRCP (UK)  FRCA
*Senior Registrar,*
*Sir Humphry Davy Department of Anaesthesia,*
*Bristol Royal Infirmary,*
*Bristol, UK*

**T. M. Craft**
MB BS  FRCA
*Consultant in Anaesthesia and Intensive Care,*
*Royal United Hospital,*
*Bath, UK*

Consultant Editor:

**P. J. F. Baskett**
MB BCh  BAO  FRCA
*Consultant Anaesthetist,*
*Frenchay Hospital and Bristol Royal Infirmary,*
*Bristol, UK*
*Chairman of the European Resuscitation Council*

*β*IOS
SCIENTIFIC
PUBLISHERS

©BIOS Scientific Publishers Limited, 1994

All rights reserved. No part of this book may be reproduced or transmitted, in any form or by any means, without permission.

First published in the United Kingdom 1994 by
BIOS Scientific Publishers Limited,
St Thomas House, Becket Street, Oxford OX1 1SJ, UK.

A CIP catalogue record for this book is available from the British Library.

ISBN 1 872748 53 8

Printed by Redwood Books Ltd, UK.

# Foreword

A patient requiring resuscitation is the single most chal-
lenging situation to face the clinician. The initial resuscitation of these patients is frequently performed by the
most junior members of staff with the least experience.
There is no substitute for practical, hands-on experience
and the authors of this book stress this aspect. There is,
however, a need for the clinician to have a compact and
easily accessible source of information in these situa-
tions. This book provides that information in a clear and
concise form. Flow diagrams and decision trees provide
a readily accessible and user-friendly approach.

The authors have had extensive experience in these sit-
uations and the problems frequently faced by junior
members of staff. The advice is clear, crisp and dogmat-
ic, and based on the most up-to-date recommendations.
It will be an invaluable *aide-memoire* for junior medical
staff, medical students, and nursing and paramedic staff
involved in acute care. Don't just keep it in your pocket.
Read it, use it and keep using it.

*Colin Robertson, FRCP FRCS FFAEM*
*Chairman*
*Resuscitation Council (UK)*

# Preface

This book contains key information for anyone involved in the resuscitation of patients. It has been collated by the authors during their practice of medicine, intensive care, and anaesthesia and includes the latest recommendations from bodies such as the European Resuscitation Council and the Resuscitation Council (UK).

Treatment of patients in an emergency situation does not afford the resuscitator the luxury of time for deliberation and discussion. It is essential, therefore, that certain data are immediately to hand.

This book includes many protocols, flow diagrams, and decision trees that have been derived from information disseminated during training courses such as those for Basic Cardiac Life Support (BCLS), Advanced Cardiac Life Support (ACLS), Advanced Trauma Life Support (ATLS), Paediatric Advanced Life Support (PALS), and Advanced Paediatric Life Support (APLS). We commend these courses and *Resuscitation: key data* is in no way designed to replace the invaluable experience that can be gained from them. The protocols should not, however, be construed as prohibiting flexibility where appropriate.

The final section of the book contains normal values for a range of investigations. Included here is ample space for the reader to make notes and record *aides-mémoires* that will result in a personalized book that is truly indispensible.

A common theme running throughout all resuscitation guidelines is the need to call for help at the earliest opportunity and this cannot be stressed enough.

*Michael J. A. Parr*
*Timothy M. Craft*

# Acknowledgements

We would like to acknowledge the generosity of the following who gave permission for their figures to be reproduced within this book:

The European Resuscitation Council (p. 2)
    *Resuscitation,* 1992;**24:**103-110.

The European Resuscitation Council (p. 5-7)
    *Resuscitation,* 1992;**24:**111-121.

Dr G Burton and Schwarz Pharma Ltd (pp. 13-14)

Dr P Oakley (pp. 54-55)

The British Medical Association (pp. 61,63)
    *ABC of Major Trauma.* London: British Medical Association, 1991.

Surgery, Gynecology & Obstetrics (p. 66)
    *Surgery, Gynecology & Obstetrics,* 1944;**79:**352-358.

# Contents

| ADULT RESUSCITATION |
|---|

## Contents

## PAEDIATRIC RESUSCITATION

**Contents**

# Abbreviations

| | |
|---|---|
| ABG | Arterial blood gas |
| BSA | Body surface area |
| CI | Cardiac index |
| CO | Cardiac output |
| CPR | Cardiopulmonary resuscitation |
| CT | Computered tomography |
| CVP | Central venous pressure |
| CXR | Chest X-ray |
| FBC | Full blood count |
| $FIO_2$ | Fractional inspired oxygen concentration |
| GCS | Glasgow coma score |
| Hct | Haematocrit |
| ICP | Intracranial pressure |
| i.m. | Intramuscular |
| IPPV | Intermittent positive pressure ventilation |
| i.v. | Intravenous |
| MAP | Mean arterial pressure |
| PEF | Peak expiratory flow |
| RBC | Red blood cell |
| U+E | Urea and electrolytes |
| VF | Ventricular fibrillation |

# Assessment of the apparently lifeless victim

Gently shake casualty's shoulders. Say in a loud voice: "Are you all right?".

If casualty is unresponsive:

1. Shout for help.

2. **ABC** (Airway, Breathing, Circulation):

   A. Open airway by loosening tight clothing around the neck and remove obvious obstruction from the mouth. With two fingertips under the point of the chin, lift the chin. An obstructed airway may also be relieved by the use of backward head-tilt and/or jaw thrust but see below.

   B. Look, listen, and feel for breathing for 5 seconds. Look for chest movements, listen at the mouth for breath sounds, feel for moving air at your cheek.

   C. Check for a carotid pulse for 5 seconds.

Follow the flow chart on page 2 according to responsiveness

**Note:**
Avoid head tilt in patients with suspected spinal injury.

| ADULT RESUSCITATION |

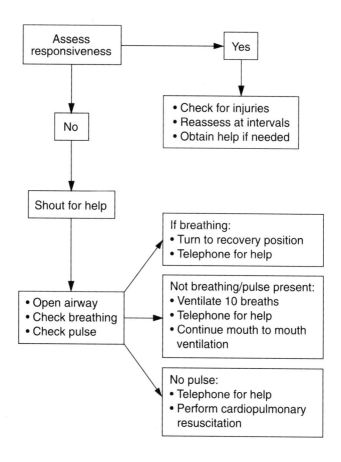

**Note:**

In the absence of a pulse, health care professionals may give a precordial thump before commencing cardiac compressions.

**ADULT RESUSCITATION**

# Basic life support

## VENTILATION

The target tidal volume for an adult is 800-1200 ml.
The target rate for a victim who is not breathing but who has a pulse is 10 breaths per minute.
Inflation should take about 2 seconds.
If the victim is not breathing and has no pulse, two ventilations should be given before commencing chest compressions and then after every 15 compressions. If two rescuers are performing CPR together a single ventilation should be given after every five compressions.

## CHEST COMPRESSIONS

Compressions are performed at the junction of the upper two-thirds and the lower third of the sternum.
The target rate of compressions for an adult is at least 80 per minute.
The sternum should be depressed by 4-5 cm during each compression.
Pressure should be applied vertically and smoothly with the same time being spent in the compressed phase as the relaxed phase.

Do not interrupt resuscitation to check for a pulse until advanced life support has commenced or until signs of life such as swallowing, limb movement, or breathing have returned.

**ADULT RESUSCITATION**

**4**
# Advanced life support

The following pages contain treatment recommendations presented as algorithms.

By their very nature, algorithms oversimplify. They are not intended to replace clinical understanding or prohibit flexibility; adopting a cookbook approach to the management of acutely ill patients does not relieve the chef of the need to think.

When applying these treatment recommendations it is essential to remember that adequate oxygenation and ventilation through a clear airway, chest compressions, and defibrillation are always more important than the administration of drugs.

The algorithms assume that the underlying rhythm being treated persists. The treatment sequence should, of course, be interrupted at any point should the arrhythmia abort.

Above all, **treat the patient, not the monitor.**

**ADULT RESUSCITATION**

# Ventricular fibrillation/pulse-less ventricular tachycardia

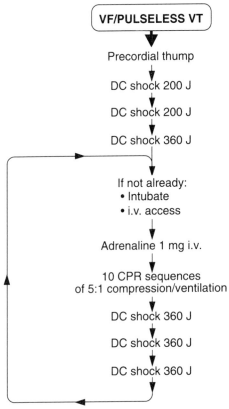

**VF/PULSELESS VT**

Precordial thump

DC shock 200 J

DC shock 200 J

DC shock 360 J

If not already:
• Intubate
• i.v. access

Adrenaline 1 mg i.v.

10 CPR sequences
of 5:1 compression/ventilation

DC shock 360 J

DC shock 360 J

DC shock 360 J

**Notes:**
Adrenaline should be given during the loop every 2-3 min.
Continue loops for as long as defibrillation is indicated.
After three loops consider an antiarrhythmic or alkalizing agent.

**ADULT RESUSCITATION**

# Asystole

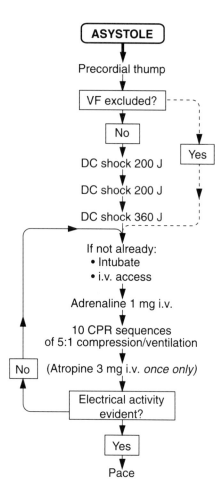

**Note:**
After three loops consider 5 mg adrenaline.

**ADULT RESUSCITATION**

# Electromechanical dissociation

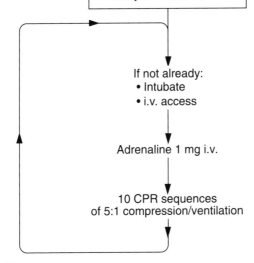

ELECTROMECHANICAL
DISSOCIATION

Think of and, if indicated,
give specific treatment for:

Hypovolaemia
Tension pneumothorax
Cardiac tamponade
Pulmonary embolism
Drug overdose/intoxication
Hypothermia
Electrolyte imbalance

If not already:
• Intubate
• i.v. access

Adrenaline 1 mg i.v.

10 CPR sequences
of 5:1 compression/ventilation

**Note:**
After three loops consider 5 mg adrenaline.

**ADULT RESUSCITATION**

# Supraventricular tachycardia

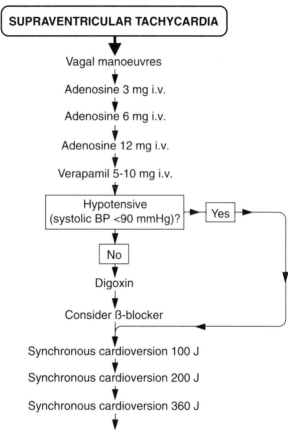

```
┌──────────────────────────────────────┐
│   SUPRAVENTRICULAR TACHYCARDIA        │
└──────────────────────────────────────┘
                  ↓
          Vagal manoeuvres
                  ↓
          Adenosine 3 mg i.v.
                  ↓
          Adenosine 6 mg i.v.
                  ↓
          Adenosine 12 mg i.v.
                  ↓
          Verapamil 5-10 mg i.v.
                  ↓
      ┌──────────────────────┐
      │     Hypotensive      │──→ ┌─────┐
      │ (systolic BP <90 mmHg)?│   │ Yes │──→
      └──────────────────────┘    └─────┘
                  ↓
              ┌──────┐
              │  No  │
              └──────┘
                  ↓
              Digoxin
                  ↓
          Consider ß-blocker
                  ↓
    Synchronous cardioversion 100 J
                  ↓
    Synchronous cardioversion 200 J
                  ↓
    Synchronous cardioversion 360 J
                  ↓
Correct underlying metabolic/electrolytic abnormalities
```

**Notes:**

Vagal manoeuvres include the Valsalva manoeuvre, and carotid sinus massage (performed unilaterally and only after a carotid bruit has been excluded).

ß-blockade after verapamil may result in AV node standstill.

**ADULT RESUSCITATION**

# Bradycardia

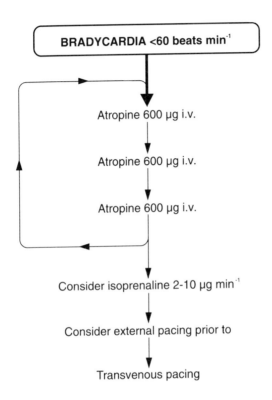

**BRADYCARDIA <60 beats min⁻¹**

Atropine 600 μg i.v.

Atropine 600 μg i.v.

Atropine 600 μg i.v.

Consider isoprenaline 2-10 μg min⁻¹

Consider external pacing prior to

Transvenous pacing

**Notes:**
Bradycardia only requires treatment in the presence of signs or symptoms such as:

    evidence of acute myocardial ischaemia or infarction
    hypotension (systolic BP <90 mmHg)
    altered mental state
    chest pain, dyspnoea.

Continue loop to a maximum dose of 3 mg of atropine.

# Hyperkalaemia

The ECG changes associated with hyperkalaemia include:

peaked T waves
loss of the P wave
wide, slurred QRS complex
ventricular tachycardia or fibrillation.

Emergency treatment of hyperkalaemia comprises:

1. Stop administration of potassium.

2. 10 ml 10% calcium chloride i.v. (extreme caution in digitalized patients).

3. 50 ml 50% glucose with 20 units of short-acting insulin followed, if necessary, by 1 litre 20% glucose with 100 units of insulin infused at 2 ml kg$^{-1}$ h$^{-1}$.

4. Promote alkalinization by hyperventilation if intubated, or give i.v. sodium bicarbonate.

**ADULT RESUSCITATION**

# Antiarrhythmic drug doses

**Digoxin:** initial i.v. dose over >1 h        0.75-1 mg.

**Lignocaine:** initial slow bolus        100 mg,
    followed by infusion of               2-4 mg min$^{-1}$.

**Amiodarone:** over 20-120 min (via     5 mg kg$^{-1}$
    central vein) max. dose in 24 h      1.2 g.

**Bretylium:** over 8-10 min repeated after  5-10 mg kg$^{-1}$
    1-2 h to a total dosage of         30 mg kg$^{-1}$.

**Adenosine:** by rapid i.v. injection     3 mg.
    2nd dose (if required)            6 mg,
    3rd dose (if required)            12 mg.
    Further dosage not recommended

**Verapamil:** by slow bolus          5-10 mg.

**Esmolol:** initial i.v. dose over 1 min    500 $\mu$g kg$^{-1}$,
    followed by infusion for 4 min of    50 $\mu$g kg$^{-1}$ min$^{-1}$.
    The loading dose may then be repeated
    and the maintenance infusion rate doubled
    if the response is inadequate.

**ADULT RESUSCITATION**

# Infusion drug doses

| | |
|---|---|
| **Dopamine** | 2-15 µg kg$^{-1}$ min$^{-1}$. |
| **Dobutamine** | 2-20 µg kg$^{-1}$ min$^{-1}$. |
| **Adrenaline** | 0.01-0.5 µg kg$^{-1}$ min$^{-1}$. |
| **Noradrenaline** | 0.01-0.5 µg kg$^{-1}$ min$^{-1}$. |
| **Isoprenaline** | 0.5-10 µg min$^{-1}$. |
| **Aminophylline:** loading dose (over 20 min) followed by | 5 mg kg$^{-1}$ 500 µg kg$^{-1}$ h$^{-1}$. |
| **Nitroprusside:** initial rate Usual dose range Maximum dose | 0.3-1 µg kg$^{-1}$ min$^{-1}$. 0.5-6 µg kg$^{-1}$ min$^{-1}$. 8 µg kg$^{-1}$ min$^{-1}$. |
| **Glyceryl trinitrate** | 10-200 µg min$^{-1}$. |

**Note:**
These doses are the usual ranges and do not preclude the use of an initial bolus or much higher infusion doses in exceptional circumstances.

**ADULT RESUSCITATION**

# Drug infusion nomograms

For solutions containing 1 mg ml$^{-1}$ a straight line is drawn from "patient dose rate" to "body weight" and the "infusion pump rate" read from the nomogram scale.

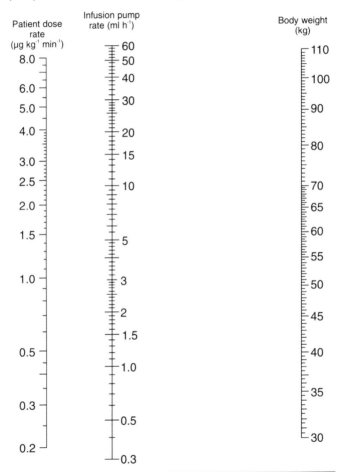

| Patient dose rate ($\mu$g kg$^{-1}$ min$^{-1}$) | Infusion pump rate (ml h$^{-1}$) | Body weight (kg) |

**ADULT RESUSCITATION**

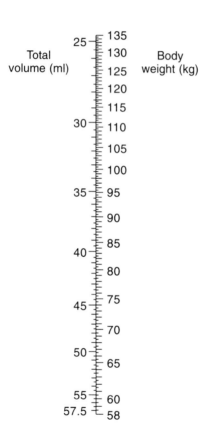

### Dopamine, dobutamine, or aminophylline

use 200 mg and make up with 5% glucose to a total volume as per the adjacent scale

1 ml h$^{-1}$ gives 1.0 µg kg$^{-1}$ min$^{-1}$

### Sodium nitroprusside

use 50 mg and make up with 5% glucose to a total volume as per the adjacent scale

1 ml h$^{-1}$ gives 0.25 µg kg$^{-1}$ min$^{-1}$

### Adrenaline, noradrenaline or isoprenaline

use 5 mg and make up with 5% glucose to a total volume as per the adjacent scale

1 ml h$^{-1}$ gives 0.025 µg kg$^{-1}$ min$^{-1}$

Total volume (ml)

Body weight (kg)

25
30
35
40
45
50
55
57.5

135
130
125
120
115
110
105
100
95
90
85
80
75
70
65
60
58

**ADULT RESUSCITATION**

# Chest pain and possible acute myocardial infarction

This algorithm suggests a line of management for patients not in hospital presenting with acute chest pain.

---

**EMERGENCY SERVICES**

Oxygen
Large bore i.v.
Monitor vital signs
Nitroglycerin sublingually
?Pain relief with opioids?
Rapid transport to Emergency Department
Notification of Emergency Department of impending arrival

---

**EMERGENCY DEPARTMENT**
Rapid triage of patients with chest pain

---

**ASSESSMENT**

**Immediate:**

Oxygen by mask (4 l min$^{-1}$)

Vital signs (ECG, BP, oxygen saturation)

Start i.v.

12 lead ECG

Brief history and examination

Decide on suitability for thrombolytic therapy

**Soon:**

CXR

Blood samples to the lab.

Referral as needed

---

**TREATMENTS TO CONSIDER IF EVIDENCE OF MYOCARDIAL INFARCTION**

Nitroglycerin sublingually if systolic BP >90 mmHg

Morphine i.v.

Thrombolytics (within 30-60 min of arrival)

Nitroglycerin i.v.i. providing systolic BP >90 mmHg

Heparin and/or aspirin

ß-blockade

PTCA (percutaneous transluminal coronary angioplasty)

---

**ADULT RESUSCITATION**

# Normal ECG values

The standard speed for recording an ECG is 25 mm sec$^{-1}$.

At this rate one large square represents 0.2 sec.

At this rate one small square represents 0.04 sec.

Every ECG should be calibrated vertically; two large squares = 1 mV.

## NORMAL ADULT VALUES

| | |
|---|---|
| Rate | 60-100 min$^{-1}$ |
| PR interval | 0.12-0.20 sec |
| P wave - maximum height: | 2.5 mm |
|       - maximum duration: | 0.11 sec |
| QRS - axis: | -30° to +90° |
|       - duration: | <0.1 sec |
| $QT_c$ (corrected for heart rate): | <0.42 sec |

# Normal cardiac and haemodynamic values

| | |
|---|---|
| MAP | 70-105 mmHg |
| CVP | 0-7 mmHg |
| MPAP | 9-16 mmHg |
| PCWP | 8-12 mmHg |
| | |
| CO | 4-8 l min$^{-1}$ |
| CI | 2.5-4 l min$^{-1}$ |
| | |
| SV | 60-130 ml beat$^{-1}$ |
| SVI | 35-70 ml beat$^{-1}$ m$^{-2}$ |
| LVSW = (MAP - PAWP) x SV x 0.0136 | 44-68 g-m m$^{-2}$ beat$^{-1}$ |
| RVSW = (MAP - RAP) x SV x 0.0136 | 4-8 g-m m$^{-2}$ beat$^{-1}$ |

$$PVR = \frac{MPAP - PAWP}{CO} \times 79.9 \qquad 25\text{-}125 \text{ dyn sec}^{-1} \text{ cm}^{-5}$$

$$SVR = \frac{MAP - RAP}{CO} \times 79.9 \qquad 960\text{-}1400 \text{ dyn sec}^{-1} \text{ cm}^{-5}$$

$$(A\text{-}a)DO_2 = \frac{}{FIO_2 \times 94.8 - PaCO_2 - PaO_2} \qquad <3 \text{ kPa}$$

**Notes:**
RAP = Right atrial pressure,
MPAP = mean pulmonary artery pressure,
PCWP = pulmonary capillary wedge pressure,
LVSW = left ventricular stroke work,
RVSW = right ventricular stroke work,
PVR = pulmonary vascular resistance,
SVR = systemic vascular resistance,
(A-a)DO$_2$ = alveolar arterial oxygen difference,
1 kPa = 7.5 mmHg.

**ADULT RESUSCITATION**

# Sequence of trauma management

1. Perform rapid visual scan and obtain history; simultaneously perform the primary survey.

2. **PRIMARY SURVEY AND START OF RESUSCITATION**

   A. **Airway** (with cervical spine control)
   Relieve airway obstruction e.g. chin lift, jaw thrust, insertion of airway, intubation, surgical airway.
   Give oxygen.
   Avoid nasal airways and/or nasogastric tubes if suspicion of base of skull fracture.

   B. **Breathing** (with ventilatory support)
   Correct inadequate ventilation.
   Drain pneumothorax/haemothorax.
   Seal open chest wound.

   C. **Circulation** (with haemorrhage control)
   Site two larger than 16G i.v. cannulae and obtain blood samples.
   Correct hypovolaemia with warm fluids.
   Initial fluid infusion of 2-3 litres of crystalloid in adults, followed by blood if the patient is still hypotensive.

   D. **Disability**
   Glasgow coma score (q.v.).
   Assess pupil responses.
   Assess limb movements.

   E. **Expose** completely for examination but avoid inducing hypothermia.

**ADULT RESUSCITATION**

## 2. PRIMARY SURVEY Cont.

Monitoring:
    ECG
    Blood pressure
    Temperature
    Oxygen saturation by pulse oximetry
    Urine output
    End expired tidal carbon dioxide (if intubated)

## 3. RESUSCITATION PHASE

Continuing management of problems as they are detected in the primary survey.

## 4. SECONDARY SURVEY

*Sequence of detailed head to toe examination:*

Scalp and head
Maxillofacial
Eyes, ears, nose and throat
Neck and cervical spine
Chest
Abdomen
Pelvis and rectum
Extremities (vascular and musculoskeletal)
Neurological function.

## 5. DEFINITIVE CARE

## ADULT RESUSCITATION

## Procedures:

Cervical and thoracolumbar immobilization
Airway insertion; oral or nasal
Intubation; oral or nasal
Cricothyroidotomy; needle or surgical
Thoracocentesis; needle or chest tube
Vascular access; peripheral, central, or surgical
Pericardiocentesis
Oro/nasogastric tube
Urinary catheter
Diagnostic peritoneal lavage
Immobilization of fractures

## Investigations:

Blood samples for:
    FBC, U&E, X-match, and ABG.
X-rays:
    lateral cervical spine from $C_1$ to $C_7$ (including the top of $T_1$)
    chest
    pelvis
    others as indicated.

## Notes:

Head and cervical spine immobilization is achieved by manual in-line stabilization (MILS) or semi-rigid collar with sand bags and tape.

Thoracolumbar immobilization is achieved with the use of a long spine board.

Any evidence of a urethral injury is an indication for retrograde urethrography and is a contraindication to perurethral bladder catheterization.

**ADULT RESUSCITATION**

# Diagnostic peritoneal lavage

Lavage is performed with 10 ml kg$^{-1}$ of lactated Ringer's solution or 0.9% saline.

The insertion site for lavage is in the midline, one-third of the distance from the umbilicus to the symphysis pubis.

## POSITIVE RESULT

Aspiration of gross blood >10 ml.
Lavage fluid exits via chest tube or urinary catheter.
Lavage fluid contains:
   Evidence of food, foreign particles, faeces or bile.
   RBC >100 000 x 10$^6$ l$^{-1}$ from blunt trauma.
   RBC >50 000 x 10$^6$ l$^{-1}$ from penetrating trauma.
   WBC >500 x 10$^6$ l$^{-1}$.

## NEGATIVE RESULT

Lavage fluid contains:
   RBC <20 000 x 10$^6$ l$^{-1}$.
   WBC <100 x 10$^6$ l$^{-1}$.

# Trauma in pregnancy

The priorities of assessment and management described earlier (pp. 19-22) also apply to pregnant victims of trauma.

Key points to remember for trauma in pregnancy:

There are two patients to consider.

Placental blood flow is compromised to maintain maternal circulation. Careful monitoring of both mother and fetus are essential for favourable fetal outcome.

Aortocaval compression is avoided by the use of left uterine displacement provided by a wedge under the right hip/spine board.

The normal hyperventilation of late pregnancy which results in a $PaCO_2$ of 4 kPa should be maintained if assisted ventilation is employed.

Diagnostic peritoneal lavage is performed in the midline above the uterus.

# Glasgow coma score

| Behaviour | Response | | Score |
|---|---|---|---|
| Eye opening | Spontaneous | E | 4 |
| | To speech | | 3 |
| | To pain | | 2 |
| | Nil | | 1 |
| Motor response (to verbal/painful stimulus) | Obeys | M | 6 |
| | Localizes | | 5 |
| | Withdraws | | 4 |
| | Abnormal flexion | | 3 |
| | Abnormal extension | | 2 |
| | Nil | | 1 |
| Verbal response | Orientated | V | 5 |
| | Confused | | 4 |
| | Inappropriate words | | 3 |
| | Incomprehensible | | 2 |
| | Nil | | 1 |

The highest score possible is 15.
The lowest score possible is 3.

# Management of severe head injury

**GCS ≤ 8** (with or without localizing signs)

Intubate following rapid sequence induction with allowance for cardiovascular status.
Hyperventilate to $PaCO_2$ 3.0-3.5 kPa.
Sedate and paralyse.
Avoid impairing cerebral venous drainage:
    i.e. avoid head down position, prevent cervical collar venous compression.
Restrict i.v. fluids.
Mannitol 1 g kg$^{-1}$ i.v. infusion over 20 min.
Consider frusemide.
Immediate CT scan.
Immediate neurosurgical consultation.

## GCS > 8

May need intubation and sedation to facilitate scan and prevent secondary brain injury.
Secondary brain injury may be caused by:
    hypoxia, hypercarbia, hypotension, increased ICP, seizures, and hyperthermia.
Urgent CT scan.
Urgent neurosurgical consultation.

# Brain death criteria

Performed by two independent doctors, one a consultant the other a consultant or senior registrar with 5 years post-registration experience.

Pre-conditions:

1. Confirm the underlying condition that resulted in irremediable brain damage
2. Exclude the presence of cerebral depressants.
3. Exclude the presence of muscle relaxants.
4. Exclude metabolic or endocrine abnormalities.
5. Exclude hypothermia. Temperature should be >35°C.

Tests are positive (for brain death) if they confirm the absence of brain stem reflexes, i.e. there are:

no pupil responses
no corneal reflexes
no vestibulo-ocular reflexes
no motor responses within cranial nerve distribution to somatic stimulation of face, limbs or trunk
no gag or cough reflexes
no ventilatory efforts in the presence of a $PaCO_2$ >6.7 kPa (insufflate oxygen).

# Initial management of a patient with burns

A. **Airway.** Give oxygen. Intubate trachea if airway in jeopardy or high level of suspicion of thermal injury to the airway.

B. **Breathing.** IPPV for ventilatory failure, following inhalation injury, impaired consciousness, or signs of airway obstruction. Victims of carbon monoxide poisoning should receive 100% oxygen and hyperbaric therapy if available.

C. **Circulation.** Insert two large bore i.v. cannulae away from burn site if possible. Take blood for U+E, FBC, and X-match. Site arterial cannula to take ABG and monitor arterial pressure directly.

Cutaneous burn results in the loss of temperature regulation. It is essential to resuscitate burned patients in a warm environment, using warmed fluids, warmed and humidified gases, and avoiding unnecessary exposure.

Control pain with bolus opioids and continue with an i.v. infusion.

Consider escharotomy or fasciotomy for tissue decompression and relief of ischaemia.

Give tetanus toxoid.

**ADULT RESUSCITATION**

# Estimation of burn size

Rapid assessment: 'Rule of nines' for adults (the body surface is divided into areas equivalent to ~9% each and the burned areas added together).

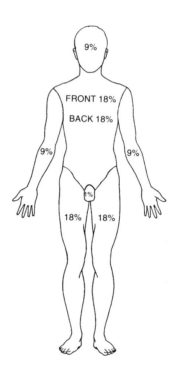

**Notes:**

The patient's palm surface area = approx. 1% BSA.

For a more accurate assessment of the surface area involved use the Lund and Browder chart (p. 66).

**ADULT RESUSCITATION**

# Fluid replacement regimens

## 1. COLLOID REQUIREMENT

The volume of colloid required from the time of the burn (millilitres to be infused per time period) is:

$$0.5 - 0.65 \text{ ml kg}^{-1} \times \text{percentage surface area of burn}$$

(e.g. 1400 - 1820 ml, for a 70 kg man with 40% burn).

This volume should be given during **each** of the following time periods:

Time in hours from time of burn
0-4, 4-8, 8-12, 12-18, 18-24, 24-36

## 2. PLASMA DEFICIT

The plasma deficit is derived from:

$$\text{Blood volume} - \left( \frac{\text{blood volume} \times \text{normal Hct}}{\text{observed Hct}} \right)$$

(e.g. 630 ml deficit for a 70 kg man with a measured Hct of 0.5).

**Notes:**
The normal blood volume of an adult is 75 ml kg$^{-1}$.
The normal Hct of an adult female is 0.4, and 0.44 for a male.

**ADULT RESUSCITATION**

## 3. BLOOD REQUIREMENT

Blood is transfused in an adult when the calculated required volume exceeds 15% of the blood volume and in a child when it exceeds 10% of the blood volume.  The volume to be replaced is calculated from:

> 1% normal blood vol. x % of full thickness burn

(e.g. 2100 ml blood required for a 70 kg man with 40% full thickness burn).

## 4. METABOLIC WATER REQUIREMENT

This is calculated as:

> $1.5 - 2$ ml $kg^{-1}$ $h^{-1}$

and is infused as 5% glucose over and above colloid requirement.

## 5. SODIUM REQUIREMENT

Expect to give 0.5 mmol $kg^{-1}$ %$burn^{-1}$ daily for the first few days.

**Notes:**
The volume of intravenous fluid required will be much greater if burns are not covered.
Formulae provide only an estimate of requirements.
Volumes of fluid administered should be adjusted according to urine output (at least 1 ml $kg^{-1}$ $h^{-1}$) and vital signs.

**ADULT RESUSCITATION**

## 6. CONSTITUENTS OF I.V. FLUIDS

**Composition of crystalloid solutions**

| Fluid | Na mmol l⁻¹ | K mmol l⁻¹ | Ca mmol l⁻¹ | Cl mmol l⁻¹ | $HCO_3$ mmol l⁻¹ | Glucose g l⁻¹ |
|---|---|---|---|---|---|---|
| 0.9% saline | 150 | | | 150 | | |
| Hartmann's solution | 131 | 5 | 2 | 111 | 29 (lactate) | |
| 5% glucose | | | | | | 50 |
| 4% glucose with 0.18% saline | 30 | | | 30 | | 40 |

# Anaphylaxis

If the cause is obvious (e.g. a drug) discontinue the exposure. Then:

**A: Airway**
100% oxygen.

**B: Breathing**
Bag and mask ventilation, intubate if serious airway obstruction or if in cardiac arrest.

**C: Circulation**
I.M. adrenaline requires an effective circulation.

| Age (years) | Volume of 1:1000 adrenaline (ml) given i.m. |
|---|---|
| <1 | 0.05 |
| 1 | 0.1 |
| 2 | 0.2 |
| 3-4 | 0.3 |
| 5 | 0.4 |
| 6-12 | 0.5 |
| Adult | 0.5 - 1.0 |

Intravenous adrenaline is preferred. The initial dose is:

50 - 100 µg (0.5-1.0 ml of 1:10 000)

Continue 100 µg boluses until hypotension and bronchospasm are corrected.

**ADULT RESUSCITATION**

An i.v. infusion of adrenaline or noradrenaline may be needed:

commence adrenaline or noradrenaline at 0.025 $\mu$g kg$^{-1}$ min$^{-1}$ (see p. 14)

give rapid i.v. fluid 10-20 ml kg$^{-1}$, preferably colloid.

Further management may include:

salbutamol, 250 $\mu$g i.v. bolus then 5-20 $\mu$g min$^{-1}$ infusion

aminophylline, 5 mg kg$^{-1}$ i.v. bolus over 20 min

hydrocortisone, 500 mg i.v.

chlorpheniramine, 20 mg i.v. diluted to 10 ml, given over 1 min

sodium bicarbonate if severe acidosis persists.

**ADULT RESUSCITATION**

# Acute severe asthma

Features of life threatening asthma include:

silent chest
cyanosis
hypoxia
inadequate ventilation
unable to perform PEF
bradycardia
hypotension
exhaustion
confusion
depressed level of consciousness
acidosis
raised or normal $PaCO_2$

Management comprises:

**A: Airway**
Give oxygen.

**B: Breathing**
Administer salbutamol continuously by nebuliser.
Give a $\beta_2$ agonist i.v., e.g. salbutamol bolus 250 μg
slowly followed by a salbutamol infusion at 3-20 μg
$min^{-1}$, but more may be needed.

Hydrocortisone i.v. 200 mg 6 hourly
(child - 100 mg 6 hourly).

Aminophylline (may be given by bolus if the patient is
not already taking theophylline) 5 mg $kg^{-1}$ over 20
min.

**ADULT RESUSCITATION**

**B: Breathing** cont.
Aminophylline maintenance infusion = 0.5 mg kg$^{-1}$h$^{-1}$
(child 6 months - 9 years -  1.0 mg kg$^{-1}$ h$^{-1}$
                    10 - 16 years -  0.8 mg kg$^{-1}$ h$^{-1}$).

Ipratropium by nebulizer
  adult 0.5 mg
  child 0.25 mg (0.125 mg if very small).

Intubate and ventilate if deteriorating, or in the presence of impending exhaustion, despite maximal treatment.

**C: Circulation**
Rehydrate with potassium-containing i.v. fluids.
NB: hypokalaemia may be severe.

Sodium bicarbonate may be needed to correct acidosis but its use will result in increased $CO_2$ production.

CXR to exclude pneumothorax or pulmonary collapse.

**ADULT RESUSCITATION**

# Hypothermia

Management is determined by the degree of hypothermia.

**Mild hypothermia** (34-36°C):
    Passive external rewarming.
    Warm blankets, warm room.

**Moderate hypothermia** (30-34°C):
    Passive rewarming.
    Active rewarming of truncal areas.
    Active internal rewarming.
        Warmed i.v. fluids (to 43°C).
        Warm humidified oxygen (42-46°C).

**Severe hypothermia** (<30°C)
    Active internal rewarming.
        Warm i.v. 0.9% saline (to 43°C).
        Warm fluid lavage (43°C) potential sites:
            peritoneum via dialysis catheter
            bladder via urinary catheter
            pleura via chest tube
            oesophagus with oesophageal rewarming tubes
            pericardium if open cardiac compression.
        Warm humidified oxygen (42-46°C).
    Ventilatory support.
    Direct circulatory rewarming by arterio-venous bypass or cardiopulmonary bypass.

**Note:**
Defibrillation and cardiac drugs are unlikely to be effective at temperatures <30°C.

# Drug overdose

Poison Information Services:

| | |
|---|---|
| Belfast | 0232 240503 |
| Birmingham | 021 554 3801 |
| Cardiff | 0222 709901 |
| Dublin | 01 379964/6 |
| Edinburgh | 031 229 2477 |
| Leeds | 0532 430715 |
| London | 071 635 9191 or 071 955 5095 |
| Newcastle | 091 232 5131 |

## GENERAL PRINCIPLES

### A. Airway
Intubation is required if the laryngeal reflexes are obtunded.

### B. Breathing
Oxygen.
IPPV for inadequate ventilation.

Consider the use of specific antagonists:
Naloxone for narcotic overdose
Flumazenil for benzodiazepine overdose.

### C. Circulation
I.V. fluids to correct hypotension.
Vasopressor agents if hypovolaemia has been corrected but hypotension persists.
If arrhythmias occur ensure that hypoxia, hypercarbia, hypovolaemia, acidosis and electrolyte imbalances have been corrected.

| ADULT RESUSCITATION |
|:---:|

Consider the use of specific antagonists:
>    Sympathomimetics in ß-blocker overdose.
>    Labetolol or esmolol for cocaine or amphetamine overdose.

**Hypothermia** (see p. 37)

**Hyperthermia**
Dantrolene (1 mg kg$^{-1}$ to a maximum of 10 mg kg$^{-1}$) is of use in drug-induced hyperthermia, e.g. following amphetamine overdose.

**Convulsions**
Correct hypoxia, hypercarbia, hypovolaemia, acidosis, metabolic (including hypoglycaemia), and electrolyte imbalance.
Treat with diazepam, 5-10 mg i.v. titrated to effect.


## PREVENTING ABSORPTION

Gastric tubes and/or induced emesis should be avoided after ingestion of corrosives.

### 1. Emesis
Ipecacuanha:
>    6-18 months old: 10 ml, older children: 15 ml,
>    30 ml for adults, followed by 200 ml of water.

Contraindications to emesis:
>    Inadequate laryngeal reflexes without the protection of tracheal intubation.
>    Ingestion of petroleum derivatives and corrosives.


## ADULT RESUSCITATION

2. **Gastric lavage**

3. **Activated charcoal**
   Prevents absorption and promotes active elimination.
   The initial dose is 50 g orally, then 25 g every 4 h.

**PROMOTION OF ELIMINATION**

1. **Forced diuresis - alkaline**
   This may be indicated for poisoning with:
      barbiturates
      salicylates
      phenoxyacetate herbicides
   and is achieved by administering the following fluids
   over 3 h.

   | |
   |---|
   | 500 ml 1.26% sodium bicarbonate<br>1000 ml 5% glucose<br>500 ml 0.9% saline + potassium chloride 20 mmol |

2. **Forced diuresis - acid**
   This may be indicated for poisoning with:
      phencyclidine
      amphetamine
      fenfluramine
   and is achieved by administering the following fluids at a
   rate of 1 l h$^{-1}$ for 4 h:

   | |
   |---|
   | 500 ml 5% glucose + 1.5 g ammonium chloride<br>500 ml 5% glucose<br>500 ml 0.9% saline |

**ADULT RESUSCITATION**

## 3. Peritoneal or haemodialysis
This may be indicated for poisoning with:
  salicylates
  barbiturates
  methanol and ethanol
  ethylene glycol
  lithium.

## 4. Haemoperfusion
This may be indicated for poisoning with:
  salicylates
  barbiturates
  meprobamate
  disopyramide
  theophylline.

## SPECIFIC ANTIDOTES

BENZODIAZEPINES

### Flumazenil
100 $\mu$g increments titrated to effect.
Usual dose range 300-600 $\mu$g (max. 2 mg).
Infusion 100-400 $\mu$g h$^{-1}$.

BETA-BLOCKERS

### Glucagon
50-150 $\mu$g kg$^{-1}$ i.v. bolus over 1 min, followed by an infusion of 1-5 mg h$^{-1}$.

### Isoprenaline
10-100 $\mu$g min$^{-1}$ i.v. infusion titrated to effect.

## ADULT RESUSCITATION

CYANIDE AND DERIVATIVES

### Dicobalt edetate
600 mg by slow i.v. bolus over 1 min, followed by a further 300 mg if no response within 1 min.

### Sodium nitrite
10 ml of 3% sodium nitrite i.v. over 3 min, followed by 25 ml of 50% sodium thiosulphate i.v. over 10 min.

DIGOXIN

### Digoxin-specific Fab antibody fragments

IRON

### Desferrioxamine
2 g in 10 ml sterile water i.m.
Gastric lavage with 2 g in 1000 ml warm water.
Leave 5 g in 50 ml water in stomach.
5 mg $kg^{-1}$ $h^{-1}$ given slowly by i.v. infusion (max 80 mg $kg^{-1}$ $day^{-1}$),
or 2 g i.m. every 12 h.

METHANOL AND ETHYLENE GLYCOL

### Ethanol
50 g orally or i.v., then 10-12 g $h^{-1}$ to maintain a level of 1-2 g $l^{-1}$ (higher rates are required for alcoholics and patients on dialysis).

**ADULT RESUSCITATION**

OPIOIDS

### Naloxone
0.4-2.4 mg i.v. repeated every few minutes to a total of 10 mg.


PARACETAMOL

### Acetylcysteine
150 mg kg$^{-1}$ in 200 ml 5% glucose i.v. over 15 min; then 50 mg kg$^{-1}$ in 500 ml 5% glucose i.v. over 4 h; then 100 mg kg$^{-1}$ in 1000 ml 5% glucose i.v. over 16 h. Total dose 300 mg kg$^{-1}$ over 20 h.
Most effective when given within 8 h of ingestion.

### Methionine
2.5 g 4 hourly for four doses (10 g over 24 h).

# Paediatric resuscitation - ABC

The following pages contain key data for the resuscitation of the newborn, infant and child.

As with adults, when resuscitating neonates, or infants and children, adequate oxygenation and ventilation through a clear airway, chest compressions, and defibrillation are always more important than the administration of drugs.

Remember:

**Call for help**

**ABC:**

    **Airway**

    **Breathing**

    **Circulation**

**PAEDIATRIC RESUSCITATION**

# Basic life support - newborn

1. Call for help from a person experience in advanced resuscitation.

2. Position the baby flat or slightly head down.

3. Administer 100% oxygen via a funnel or ventilate with a face mask if apnoeic.

4. Commence external cardiac compressions if the heart rate is less than 60 min$^{-1}$ or if pulses poor or absent. Compressions are performed using both thumbs. The neonate's chest is held between both hands and the thumbs placed over the junction of the middle and lower thirds of the sternum. The sternum is depressed smoothly and rhythmically by 1-1.5 cm.
   The target rate of compressions is at least 100 min$^{-1}$.

5. Prevent heat loss. Dry the baby. Use warmed bedding and resuscitation surfaces. Keep the baby out of draughts. Resuscitate the baby under a radiant heat source. Maintain the ambient temperature at 24-25°C

6. Neonates have a high glucose need and low glycogen store. During periods of stress the baby may become hypoglycaemic. Documented hypoglycaemia should be treated with an infusion of glucose.

Observe for:
   a) symmetrical chest movement
   b) onset of regular breathing
   c) improvement in heart rate, peripheral perfusion, colour, movement, and tone.

7. Consider administering naloxone.

**PAEDIATRIC RESUSCITATION**

# Acute perinatal blood loss

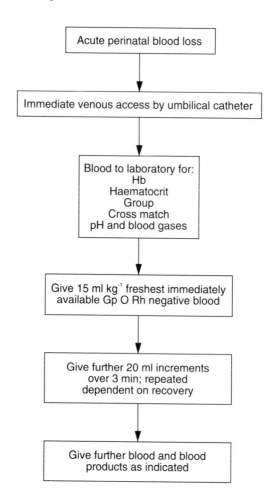

Acute perinatal blood loss

↓

Immediate venous access by umbilical catheter

↓

Blood to laboratory for:
Hb
Haematocrit
Group
Cross match
pH and blood gases

↓

Give 15 ml kg$^{-1}$ freshest immediately available Gp O Rh negative blood

↓

Give further 20 ml increments over 3 min; repeated dependent on recovery

↓

Give further blood and blood products as indicated

**PAEDIATRIC RESUSCITATION**

# Unanticipated meconium at delivery

1. Suck out the mouth and nose as soon as the head is delivered.

2. Call for help from a person experience in advanced resuscitation.

3. Place the baby flat or slightly head down.

4. Suck out mouth and nose again.

5. Give oxygen by funnel.

6. If meconium aspiration is suspected proceed to direct laryngoscopy (see p. 50).

**PAEDIATRIC RESUSCITATION**

# Meconium aspiration

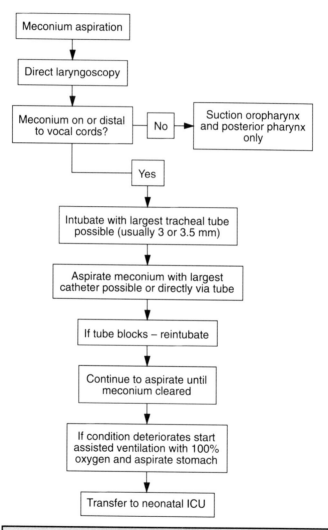

**PAEDIATRIC RESUSCITATION**

# Apgar score

|  | 0 | 1 | 2 |
|---|---|---|---|
| **Heart rate** | Absent | Slow (<100) | >100 |
| **Respiratory effort** | Absent | Weak cry, hypoventilating | Crying lustily |
| **Muscle tone** | Flaccid | Some flexion of limbs | Well flexed |
| **Colour** | Blue or white | Blue hands or feet | Healthy pink |
| **Reflex irritability** | No response | Some move- ment | Active movement |

**Note:**
The Apgar score is assessed at 1 and 5 minutes after birth.
The maximum score is 10.

**PAEDIATRIC RESUSCITATION**

# Basic life support - infants and children

The following guidelines apply to infants and children under the age of 8 years. Large or older children should be resuscitated as described for adult victims (p. 3).

## VENTILATION

The target tidal volume for a child is 7-10 ml kg$^{-1}$.
The target rate for a victim who is not breathing but who has a pulse is 20 breaths per minute.
Inflation should take about 1.5 seconds.
If the victim is not breathing and has no pulse, two ventilations should be given before commencing chest compressions and then one breath after every five compressions.

## CHEST COMPRESSIONS

Compressions are performed at the junction of the upper two-thirds and the lower third of the sternum.
The target rate of compressions for a child is at least 100 per minute.
The sternum should be depressed by 2.5-4 cm during each compression.
Pressure should be applied vertically and smoothly with the same time being spent in the compressed phase as the relaxed phase.

Do not interrupt resuscitation to check for a pulse until advanced life support has commenced or until signs of life such as swallowing, limb movement, or breathing have returned.

**PAEDIATRIC RESUSCITATION**

# Paediatric resuscitation chart

Tracheal tube

| Oral length (cm) | Internal diameter (mm) |
|---|---|
| 18-21 | 7.5-8.0 (cuffed) |
| 18 | 7.0 (uncuffed) |
| 17 | 6.5 |
| 16 | 6.0 |
| 15 | 5.5 |
| 14 | 5.0 |
| 13 | 4.5 |
| 12 | 4.0 |
| | 3.5 |
| 10 | 3.0-3.5 |

Length (cm) — Age (years) vs Weight (kg) graph

| | 0.5 | 1 | 2 | 3 | 4 | 5 |
|---|---|---|---|---|---|---|
| Adrenaline (ml of 1 in 10 000) *initial* <br> intravenous or intraosseous | 0.5 | 1 | 2 | 3 | 4 | 5 |
| Adrenaline (ml of 1 in 1000) *subsequent* <br> intravenous or intraosseous (or initial endotracheal) | 0.5 | 1 | 2 | 3 | 4 | 5 |
| Atropine (ml of 100 µg ml⁻¹) <br> intravenous or intraosseous (or double if endotracheal) | 1 | 2 | 4 | 6 | 6 | 6 |
| Atropine (ml of 600 µg ml⁻¹) | - | 0.3 | 0.7 | 1 | 1 | 1 |
| Bicarbonate (ml of 8.4%) <br> intravenous or intraosseous (dilute to 4.2% in infants) | 5 | 10 | 20 | 30 | 40 | 50 |
| Calcium chloride (ml of 10%) <br> intravenous or intraosseous | 0.5 | 1 | 2 | 3 | 4 | 5 |
| Diazepam (ml of 5 mg ml⁻¹ emulsion) <br> intravenous or rectal | 0.4 | 0.8 | 1.6 | 2 | 2 | 2 |
| Diazepam (mg rectal tube solution) <br> rectal | 2.5 | 5 | 10 | 10 | 10 | 10 |

**PAEDIATRIC RESUSCITATION**

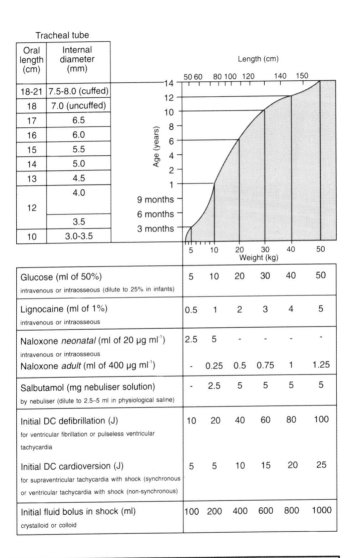

Tracheal tube

| Oral length (cm) | Internal diameter (mm) |
|---|---|
| 18-21 | 7.5-8.0 (cuffed) |
| 18 | 7.0 (uncuffed) |
| 17 | 6.5 |
| 16 | 6.0 |
| 15 | 5.5 |
| 14 | 5.0 |
| 13 | 4.5 |
| 12 | 4.0 |
| 12 | 3.5 |
| 10 | 3.0-3.5 |

| | | | | | | |
|---|---|---|---|---|---|---|
| Glucose (ml of 50%)<br>intravenous or intraosseous (dilute to 25% in infants) | 5 | 10 | 20 | 30 | 40 | 50 |
| Lignocaine (ml of 1%)<br>intravenous or intraosseous | 0.5 | 1 | 2 | 3 | 4 | 5 |
| Naloxone *neonatal* (ml of 20 µg ml⁻¹)<br>intravenous or intraosseous | 2.5 | 5 | - | - | - | - |
| Naloxone *adult* (ml of 400 µg ml⁻¹) | - | 0.25 | 0.5 | 0.75 | 1 | 1.25 |
| Salbutamol (mg nebuliser solution)<br>by nebuliser (dilute to 2.5–5 ml in physiological saline) | - | 2.5 | 5 | 5 | 5 | 5 |
| Initial DC defibrillation (J)<br>for ventricular fibrillation or pulseless ventricular tachycardia | 10 | 20 | 40 | 60 | 80 | 100 |
| Initial DC cardioversion (J)<br>for supraventricular tachycardia with shock (synchronous or ventricular tachycardia with shock (non-synchronous) | 5 | 5 | 10 | 15 | 20 | 25 |
| Initial fluid bolus in shock (ml)<br>crystalloid or colloid | 100 | 200 | 400 | 600 | 800 | 1000 |

## PAEDIATRIC RESUSCITATION

**Notes:**
Non-standard drug concentrations may be available for some of the agents.

**Atropine:** Use 100 µg ml⁻¹ by diluting 1 mg to 10 ml or 600 µg to 6 ml in 0.9% saline.

**Calcium:** 1 ml of calcium chloride 10% is equivalent to 3 ml calcium gluconate.

**Lignocaine:** Use 1% or give twice the volume of 0.5% or half the volume of 2%, or dilute appropriately.

**Salbutamol:** may also be given by slow i.v. infusion but beware the different concentrations available (e.g. 50 and 500 µg ml⁻¹).

# Paediatric trauma - principles

The same priorities of assessment and management that apply to adults also apply to victims of paediatric trauma (p. 19).

Key points to remember in paediatric trauma are:

Children often show cardiorespiratory compensation until precipitous collapse.

High surface area to body mass ratio results in an increased potential for heat loss. It is thus essential to resuscitate children in a warm environment, using warmed fluids, warmed and humidified gases, and avoiding unnecessary exposure.

The high compliance of the paediatric skeleton results in internal organ injury in the absence of obvious overlying surface injury.

If an emergency surgical airway becomes necessary, needle cricothyroidotomy and jet insufflation is preferred to surgical cricothyroidotomy because of the risk of laryngeal injury with the latter.

CT scanning offers an alternative to diagnostic peritoneal lavage for assessing abdominal trauma provided it is immediately available and does not interrupt the resuscitative process.

Failure to cannulate a vein in shocked children aged less than 6 years after two attempts is an indication to proceed to intraosseous infusion.

**PAEDIATRIC RESUSCITATION**

# Paediatric vital signs

An estimate of a child's normal systolic blood pressure can be made using the formula:

$$\text{blood pressure (mmHg)} = 80 + (\text{age} \times 2)$$

Children have a greater circulatory physiological reserve than adults. By comparison with adults, signs of impending circulatory collapse may be only slight despite considerable blood loss. The diagnosis of shock, or its precursors, in children is made on the appearance of the skin, the capillary refill and temperature of the extremities, and the presence of an altered cerebral state. Fluid resuscitation should not be withheld until the vital signs are abnormal.

|  | Age | | |
|---|---|---|---|
|  | <1 year | 2-5 years | 5-12 years |
| **Heart rate** (beats min⁻¹) | 120-140 | 100-120 | 80-100 |
| **Blood pressure** (systolic) (mmHg) | 70-90 | 80-90 | 90-110 |
| **Respiratory rate** (breaths min⁻¹) | 30-40 | 20-30 | 15-20 |
| **Blood volume** (ml kg⁻¹) | 90 | 80 | 80 |

**PAEDIATRIC RESUSCITATION**

# Paediatric resuscitation formulae

The following formulae provide rapid estimates.

### Body weight (kg)

| | |
|---|---|
| 1-8 years: | (Age x 2) + 9 |
| 8-13 years: | Age x 3 |

### Tracheal tube internal diameter (mm)

(Age/4) + 4.5

This is approximately the same size as the child's nostril or little finger.

### Tracheal tube length (cm)

| | |
|---|---|
| Oral: | (Age/2) + 12 |
| Nasal: | (Age/2) + 15 |

### Maintenance fluid requirement (per kg body weight)

| | |
|---|---|
| 0-10 kg: | $4 \text{ ml kg}^{-1} \text{ h}^{-1}$ |
| 11-20 kg: | $4 \text{ ml kg}^{-1} \text{ h}^{-1}$ for the first 10 kg<br>+ $2 \text{ ml kg}^{-1} \text{ h}^{-1}$ for the remainder |
| >20 kg: | $4 \text{ ml kg}^{-1} \text{ h}^{-1}$ for the first 10 kg<br>+ $2 \text{ ml kg}^{-1} \text{ h}^{-1}$ for the second 10 kg<br>+ $1 \text{ ml kg}^{-1} \text{ h}^{-1}$ for the remainder |

**PAEDIATRIC RESUSCITATION**

# Intraosseous infusion

This is a suitable means of administering fluid to children aged less than 6 years in whom venous cannulation has failed.

A 16 or 18G bone marrow needle is inserted into the tibia on its anteromedial surface 1-3 cm below the tibial tubercle. The needle is directed inferiorly to avoid the epiphyseal plate. In the presence of tibial fractures the distal femur may be used.

The following have been successfully administered by the intraosseous route:

Crystalloid fluids
Colloid fluids
Blood
Adrenaline
Antibiotics
Atracurium
Atropine
Calcium
Dobutamine
Dopamine
Digoxin
Glucose
Lignocaine
Midazolam
Sodium bicarbonate
Suxamethonium

# Paediatric fluid administration

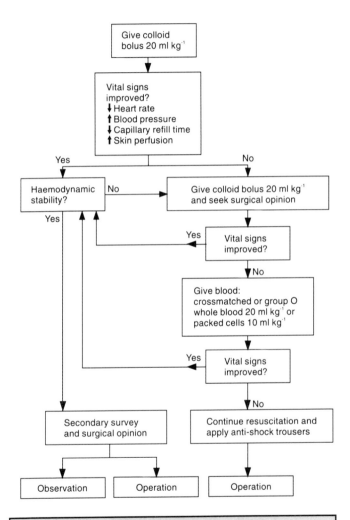

**PAEDIATRIC RESUSCITATION**

# Paediatric trauma score

|  | +2 | +1 | -1 |
|---|---|---|---|
| **Weight** | >20 kg | 10-20 kg | <10 kg |
| **Airway** | Normal | Oral or nasal airway | Intubated |
| **Blood pressure** | >90 mmHg | 50-90 mmHg | <50 mmHg |
| **Level of consciousness** | Completely awake | Obtunded or any LOC* | Comatose |
| **Open wound** | None | Minor | Major or penetrating |
| **Fractures** | None | Minor | Open or multiple |

*LOC = loss of consciousness.

**Notes:**
The paediatric trauma score (PTS) assesses child size, airway, systolic blood pressure, level of consciousness, cutaneous and skeletal injury.
Children with a PTS <8 should be managed in institutions with the facilities and experience to deal with major paediatric trauma. Children with scores >8 have the highest potential for preventable morbidity and mortality and require close observation and monitoring.

**PAEDIATRIC RESUSCITATION**

# Paediatric Glasgow coma score

| | | >1 year | <1 year | |
|---|---|---|---|---|
| **Eye opening** | 4 | Spontaneously | Spontaneously | |
| | 3 | To verbal command | To shout | |
| | 2 | To pain | To pain | |
| | 1 | No response | No response | |
| **Best motor response** | 5 | Obeys commands | | |
| | 4 | Localises pain | Localises pain | |
| | 3 | Flexion to pain | Flexion to pain | |
| | 2 | Extension to pain | Extension to pain | |
| | 1 | No response | No response | |
| | | **>5 years** | **2-5 years** | **0-2 years** |
| **Best verbal response** | 5 | Orientated and converses | Appropriate words and phrases | Smiles and cries appropriately |
| | 4 | Disorientated and converses | Inappropriate words | Cries |
| | 3 | Inappropriate words | Cries | Inappropriate crying |
| | 2 | Incomprehensible sounds | Grunting | Grunting |
| | 1 | No response | No response | No response |

**Normal aggregate score:**

| | | | |
|---|---|---|---|
| <6 months | 12 | 2-5 years | 14 |
| 6-12 months | 12 | >5 years | 14 |
| 1-2 years | 13 | | |

**PAEDIATRIC RESUSCITATION**

# Initial management of a paediatric patient with burns

The same general principles apply to the management of a child with burns as those used in the treatment of burned adults (p. 27). However, a more accurate assessment of the percentage body surface area involved is required in children and the method described by Lund and Browder may be used (p. 66).

The adequacy of fluid replacement is monitored by ensuring a urine output of at least 1 ml kg$^{-1}$ h$^{-1}$ in children aged over 1 year and at least 2 ml kg$^{-1}$ h$^{-1}$ in those under 1.

Because of the small diameter of the paediatric airway, any evidence of airway injury should result in a rapid respiratory assessment and a low threshold for early tracheal intubation.

Cutaneous burn results in the loss of temperature regulation. This combined with the high surface area to body mass ratio of a child results in an increased potential for heat loss. It is thus essential to resuscitate burned children in a warm environment, using warmed fluids, warmed and humidified gases, and avoiding unnecessary exposure.

**PAEDIATRIC RESUSCITATION**

# Estimation of burn size

Lund and Browder chart

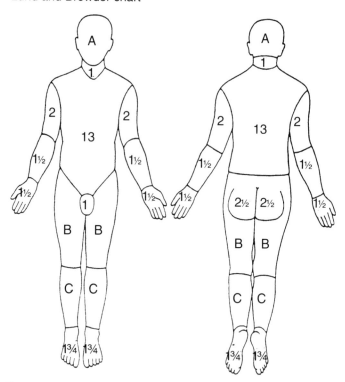

Relative percentage of body surface area affected by growth

| Area | Age (years) | | | | | |
|---|---|---|---|---|---|---|
| | 0 | 1 | 5 | 10 | 15 | Adult |
| A = ½ of head | 9½ | 8½ | 6½ | 5½ | 4½ | 3½ |
| B = ½ of one thigh | 2¾ | 3¼ | 4 | 4½ | 4½ | 4¾ |
| C = ½ of one leg | 2½ | 2½ | 2¾ | 3 | 3¼ | 3½ |

## PAEDIATRIC RESUSCITATION

# Haematology

|  | Neonate | Child | Adult |
|---|---|---|---|
| **Hb** (g dl$^{-1}$) | 18-19 | 11-14 | 13.5-17.5 (M)<br>11.5-15.5 (F) |
| **Hct** | 0.55-0.65 | 0.36-0.42 | 0.4-0.45 (M)<br>0.36-0.44(F) |
| **MCV** (fl) | 100-125 | 80-96 | 83-96 |
| **MCHC** (g l$^{-1}$) | 30-34 | 32-36 | 32-36 |
| **WBC** (x10$^9$ l$^{-1}$) | 6-15 | 5-14 | 4-11 |
| **Neutrophils**<br>(% of WBC) | 30-50 | 40-70 | 50-75 |

**Note:**
Equivalent umbilical cord blood values are:
Hb 13.5-20 g dl$^{-1}$,  Hct 0.5-0.56,  MCV 110-128 fl,
MCHC 29.5-33.5 g l$^{-1}$,  WBC 9-30 x10$^9$ l$^{-1}$,  neutrophils 50-80%.

**NORMAL VALUES**

# Coagulation

| | |
|---|---|
| **Platelet count** ($\times 10^9$ $l^{-1}$) | 150-400 |
| **Bleeding time** (min) | <7 |
| **Prothrombin time** (sec) | 11.5-15 |
| **Activated partial thrombo-plastin time** (sec) | 25-37 |
| **Thrombin time** (sec) | 10 |
| **Fibrinogen** (g $l^{-1}$) | 2-4.5 |
| **FDPs** (mg $l^{-1}$) | <10 |

**NORMAL VALUES**

# Biochemistry

| | Neonate | Child | Adult |
|---|---|---|---|
| **Na** (mmol $l^{-1}$) | 130-145 | 132-145 | 133-143 |
| **K** (mmol $l^{-1}$) | 4.0-7.0 | 3.5-5.5 | 3.6-4.6 |
| **Cl** (mmol $l^{-1}$) | 95-110 | 95-110 | 95-105 |
| **Cr** ($\mu$mol $l^{-1}$) | 28-60 | 30-80 | 60-100 |
| **Urea** (mmol $l^{-1}$) | 1.0-5.0 | 2.5-6.5 | 3-7 |
| **Mg** (mmol $l^{-1}$) | 0.6-1.0 | 0.6-1.0 | 0.7-1 |
| **Ca** (mmol $l^{-1}$) | 1.8-2.8 | 2.15-2.7 | 2.25-2.7 |
| **Phosphate** (mmol $l^{-1}$) | 1.3-3.0 | 1.0-1.8 | 0.85-1.4 |
| **Bilirubin** ($\mu$mol $l^{-1}$) | <200 | <15 | <17 |
| **Alkaline phosphatase** (U $l^{-1}$) | 150-600 | 250-1000 | 21-120 |
| **AST** (U $l^{-1}$) | <100 | <50 | 6-35 |
| **Total protein** (g $l^{-1}$) | 45-75 | 60-80 | 62-80 |
| **Albumin** (g $l^{-1}$) | 24-48 | 30-50 | 35-55 |
| **Globulin** (g $l^{-1}$) | 20-30 | 20-30 | 22-36 |

**NORMAL VALUES**

# Blood gases

**ARTERIAL**

| | |
|---|---|
| **pH** | 7.34-7.36 |
| **$PaO_2$** (kPa [mmHg]) | 12-14.67 [75-100] |
| **$PaCO_2$** (kPa [mmHg]) | 4.5-6.1 [33-46] |
| **Actual bicarbonate** (mmol $l^{-1}$) | 22-26 |
| **Standard bicarbonate** (mmol $l^{-1}$) | 22-26 |
| **Base excess** | ± 2 |
| **Oxygen saturation** | 0.96-1.0 |

**MIXED VENOUS**

| | |
|---|---|
| **pH** | 7.32-7.42 |
| **$PaO_2$** (kPa [mmHg]) | 4.96-5.6 [36-42] |
| **$PaCO_2$** (kPa [mmHg]) | 5.3-6.9 [40-52] |
| **Oxygen saturation** | 0.7-0.8 |

NORMAL VALUES

# Conversion factors

1 mmHg = 133.3 Pa = 1.36 $cmH_2O$ = 1.25 cmblood

100 mmHg = 13.3 kPa

760 mmHg = 101.3 kPa

1 kPa = 7.5 mmHg = 10.2 $cmH_2O$

1 $mmH_2O$ = 0.073 mmHg

100 kPa = 15 psi

1 atm = 1 Bar = 101.3 kPa = 1033 $cmH_2O$

**NORMAL VALUES**

# Bibliography

Advanced Life Support Working Party of the European Resuscitation Council, 1992. Guidelines for advanced life support. *Resuscitation,* 1992;**24:**111-121.

Basic Life Support Working Party of the European Resuscitation Council, 1992. Guidelines for basic life support. *Resuscitation,* 1992;**24:**103-110.

Baskett P J F. *Resuscitation Handbook (2nd ed.).* London: Mosby Year Book, 1993.

Berden H J J M, Willems F F, Hendrick J M A, Pijls N H J, Howie P W. How frequently should cardiopulmonary resuscitation training be repeated to maintain adequate skills? *British Medical Journal,* 1993;**306:**1576-1577.

Conference of the Medical Royal Colleges and their Faculties. Diagnosis of brain death. *British Medical Journal,* 1976;**2:**1187-1188.

Grande C M, ed. *Textbook of Trauma Anesthesia and Critical Care.* St Louis: Mosby Year Book, 1993.

Henry J, Volans G, eds. *ABC of Poisoning.* London: British Medical Association, 1984.

Lloyd-Thomas A R. Paediatric trauma: primary survey and resuscitation - II. *British Medical Journal,* 1990;**301:**380-382.

Lund C C, Browder N C. The estimation of areas of burns. *Surgery, Gynecology & Obstetrics,* 1944;**79:**352-358.

Recommendations of the 1992 National Conference. Guidelines for cardiopulmonary resuscitation and emergency cardiac care. *The Journal of the American Medical Association,* 1992;**268:**2171-2302.

Skinner D, Driscoll P, Earlam R, eds. *ABC of Major Trauma.* London: British Medical Association, 1991.

Wardrope J, Morris F. European guidelines on resuscitation. *British Medical Journal,* 1993;**306:**1555-1556.

# Index